Let's explore CLIMATE CHANGE together

Mosquitoes on the Rise

by Sally-Ann and Sirion Robertson

Room to Read seeks to transform the lives of millions of children in low-income communities by focusing on literacy and gender equality in education. Working in collaboration with local communities, partner organizations and governments, we develop literacy skills and a habit of reading among primary school children, and support girls to complete secondary school with the relevant life skills to succeed in school and beyond. Learn more at www.roomtoread.org.

Mosquitos on the Rise
Authors: Sally-Ann and Sirion Robertson
Editors: Naomi Mositsa, Laura Atkins
Designer: Christy Hale
ISBN 979-8-4000-0049-2

Printed in Canada

Room to Read
465 California Street, Suite 1000
San Francisco, CA. 94104
USA

Room to Read

Table of Contents

1 Hello!

Hello, friends!
Let's go out into the warm summer evening.
Can you feel the moist air on your skin?
But wait! What's that? Oh, no! A mosquito!

Mosquitoes live in many parts of the world, including South Africa.
Most mosquitoes are just an annoyance.
But some can be dangerous.
Have you ever been bitten by a mosquito?
Ouch!

2

Mosquitoes and Malaria

Do you know anyone who has had **malaria**? Mosquitoes can pass this dangerous disease on to humans.

There are many different kinds of mosquitoes. Not all of them carry malaria. Only some female **Anopheles mosquitoes** carry malaria parasites.

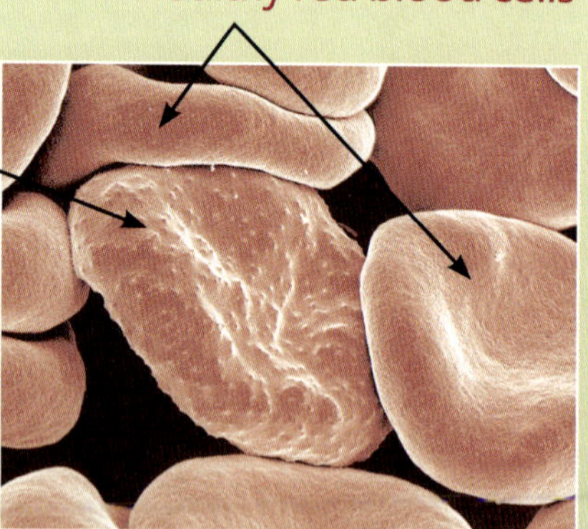

A **parasite** is an animal or plant that lives on or inside another animal or plant and gets its food from it. Some parasites are harmless. Others, such as the malaria parasite, are harmful.

Malaria is a serious disease. It makes a person very sick. If it isn't diagnosed early, and treated correctly, the person can even die from it. The main malaria symptoms are:

- High temperature ('fever')
- Cold shivers
- Hot sweats
- Headache
- Vomiting
- Diarrhoea
- Aching muscles

MALARIA TRANSMISSION CYCLE

1. If a mosquito carrying malaria parasites bites a person, malaria parasites will be injected into that person's bloodstream.

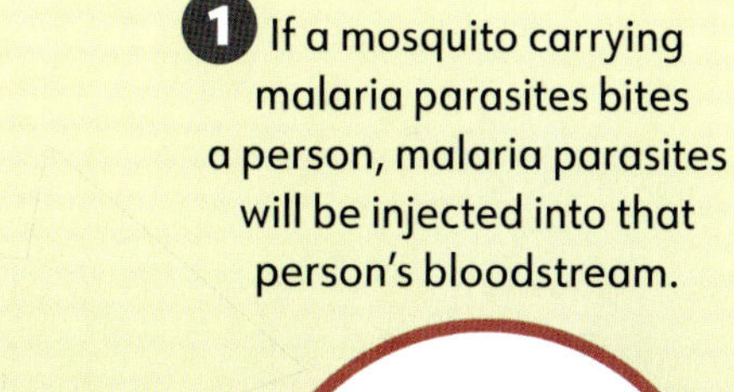

2. The malaria parasites multiply inside the person's body. The person becomes sick.

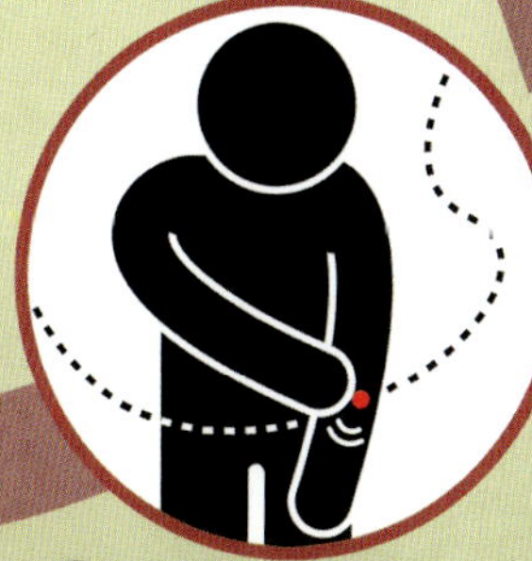

3. Another mosquito bites the sick person and sucks up some of the malaria parasites.

4. Another mosquito bites the sick person and sucks up some of the malaria parasites which multiply inside the mosquito.

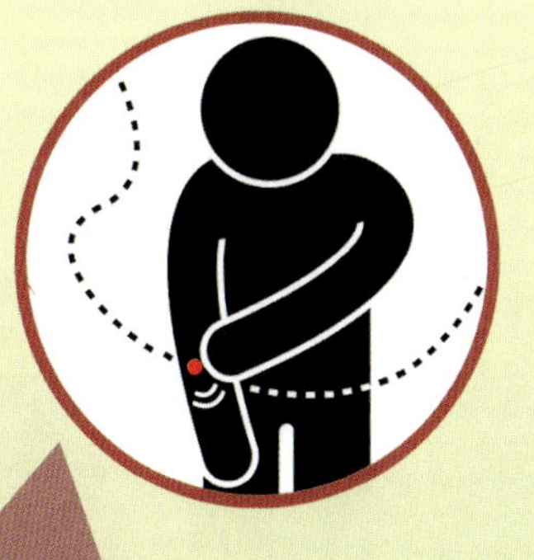

Every year more than two hundred million people get infected with malaria. That is a huge number! Nearly half a million people, including many children, die from malaria every year. That's more than a thousand people every single day. Most of the world's malaria infections happen in Africa.

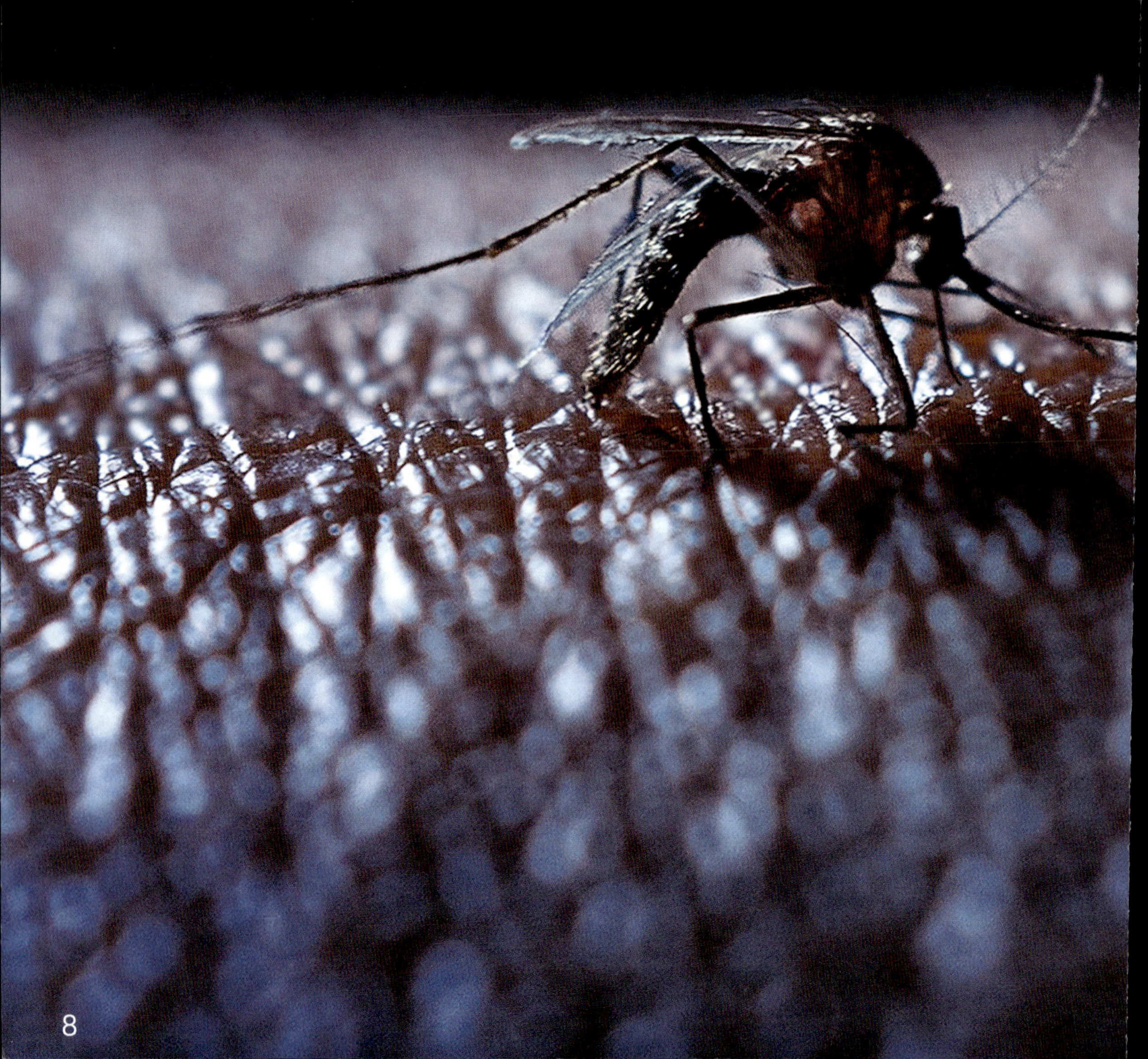

Malaria is **transmitted** more easily in hot places where there is plenty of water. This is because Anopheles mosquitoes bite more often and can complete their life cycle more quickly in these sorts of places.

A mosquito's life cycle consists of four stages. The first three stages need water. The cycle of changing from egg to larva to pupa to adult is called '**metamorphosis**'.

A Hot Wet Climate

The warmer and wetter parts of our planet are those closest to the **equator**.

Some places near the equator are hot and very dry, such as Africa's Sahara Desert. Other places, such as many countries in sub-Saharan Africa, are also hot, but have plenty of rain. Mosquitoes **thrive** in these sorts of hot and humid environments.

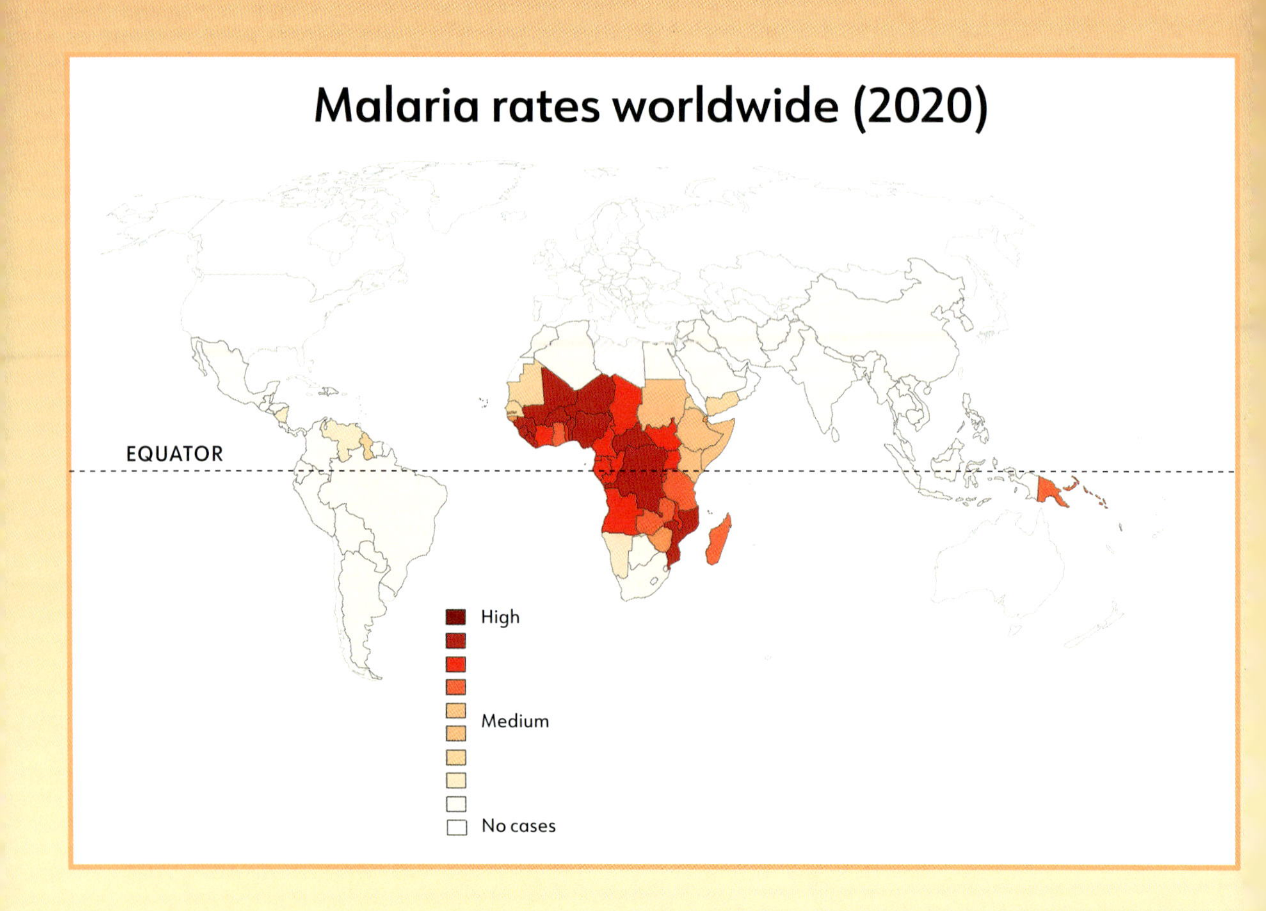

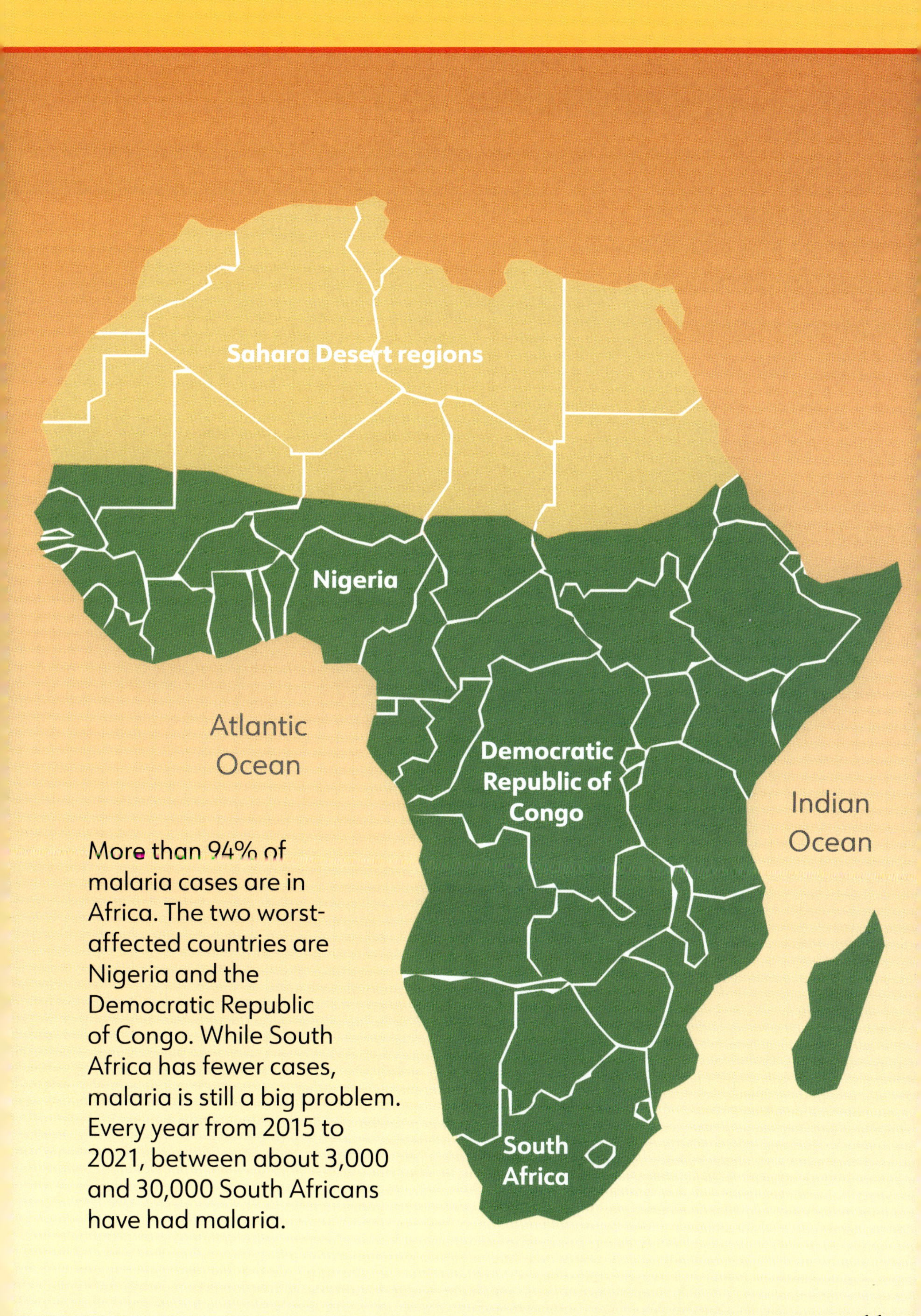

More than 94% of malaria cases are in Africa. The two worst-affected countries are Nigeria and the Democratic Republic of Congo. While South Africa has fewer cases, malaria is still a big problem. Every year from 2015 to 2021, between about 3,000 and 30,000 South Africans have had malaria.

4 Limpopo Province and Malaria

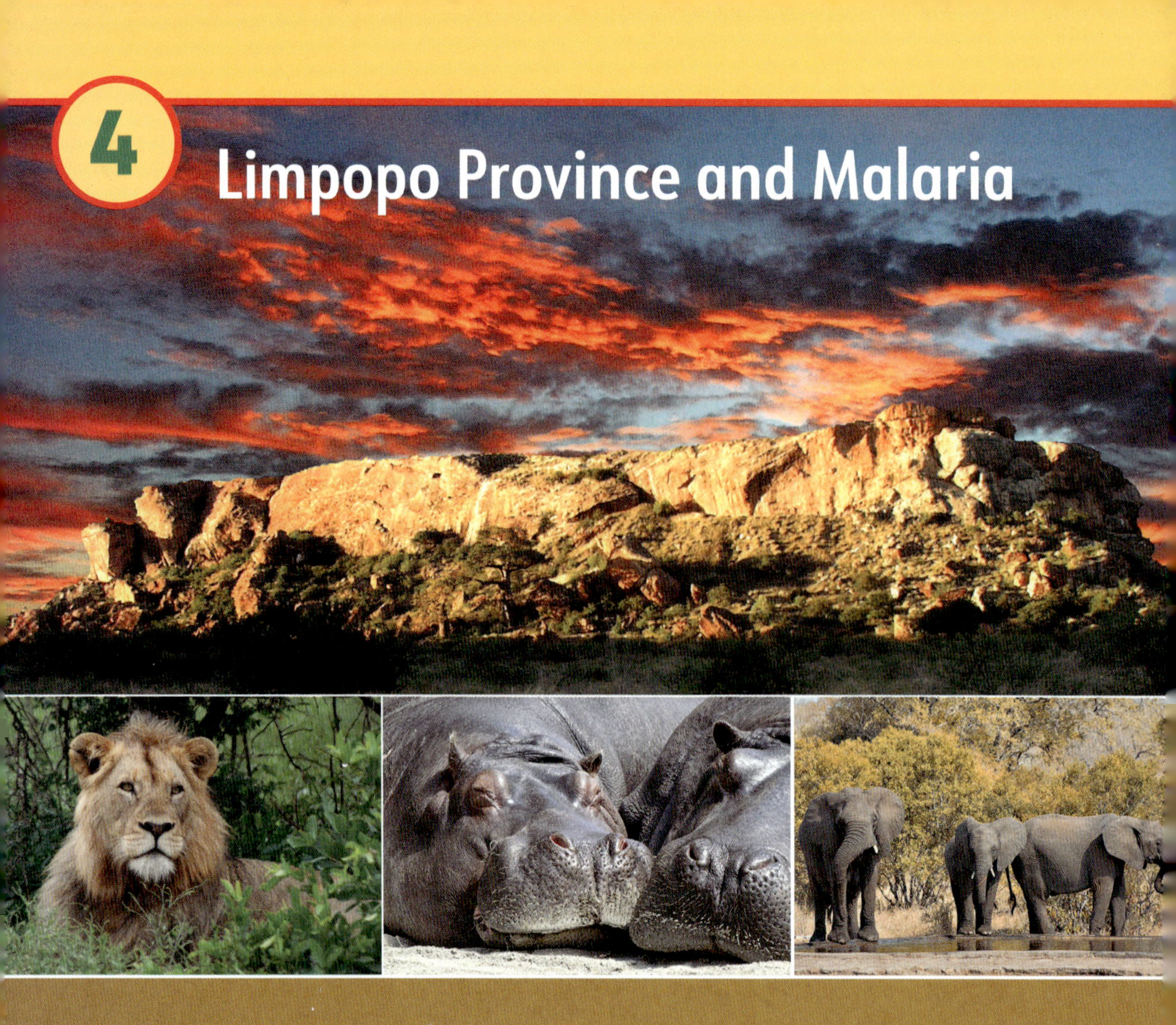

The beautiful Limpopo Province is one of the warmer and wetter regions of South Africa. The mighty Limpopo River, one of Africa's longest rivers, flows through this province.

Kruger Park, South Africa's biggest national park, lies on Limpopo's eastern border. Here many magnificent African animals roam freely. The beautiful and strange-looking baobab tree is also found here.

Some people call baobabs 'upside-down trees'!

Thousands of tourists visit Limpopo every year to enjoy its wonderful sights and sounds and its sunny climate. Many of the people living in Limpopo earn their livings from tourism.

But a shadow hangs over Limpopo. It is South Africa's main malaria "**hotspot**".

Limpopo Province has the highest malaria rates in South Africa, with thousands of cases each year. The climate creates a perfect place for mosquitoes to multiply.

Medical staff at Limpopo's many clinics and hospitals are trained to diagnose and treat malaria. It is very important that treatment starts early before people become too ill.

One of the most important actions people can take is spraying their houses with insecticides designed to **eliminate** mosquitoes.

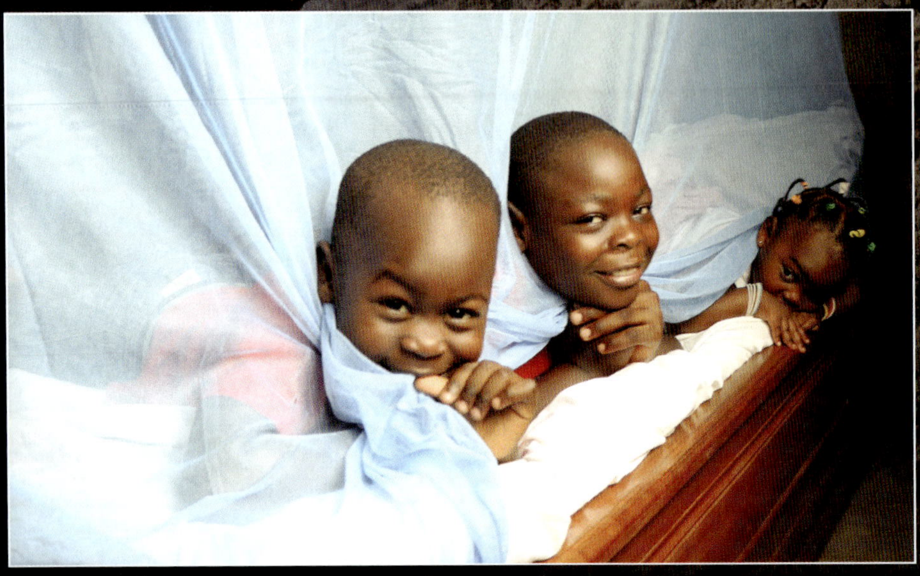

Children peeping out from under a mosquito net.
They will need to tuck up securely underneath it to sleep.

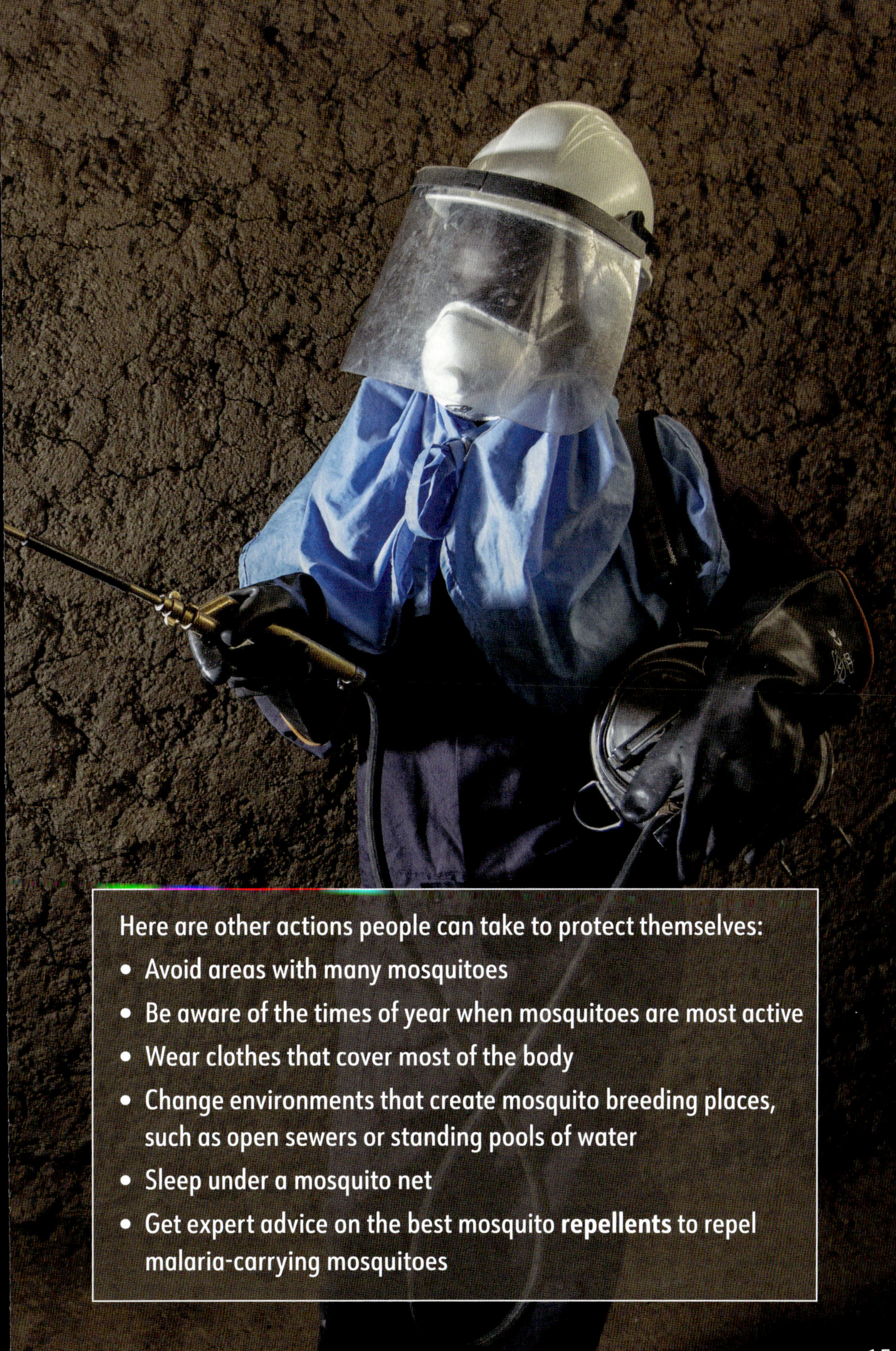

Here are other actions people can take to protect themselves:

- Avoid areas with many mosquitoes
- Be aware of the times of year when mosquitoes are most active
- Wear clothes that cover most of the body
- Change environments that create mosquito breeding places, such as open sewers or standing pools of water
- Sleep under a mosquito net
- Get expert advice on the best mosquito **repellents** to repel malaria-carrying mosquitoes

Rising Numbers

Limpopo Province was already South Africa's malaria hotspot, but things may be getting worse. Nearly 31,000 people got malaria in 2017. This was more than twenty times the number from the year before.

Scientists don't know for sure what caused this huge spike. One reason could be changes to Limpopo's weather patterns. Average temperatures have increased by more than 1°C over the past seven decades. The summers are hotter and drier, and the winters are warmer and wetter. These changes **extend** the time for mosquitoes to spread malaria.

Stretches of still water make perfect places for mosquitoes to lay their eggs

When weather patterns change, malaria rates change

Less rain in summer means that rivers don't flow as quickly. In still water, the eggs can hatch and the larvae can metamorphose into adults without disturbance. Mosquitoes are usually much less active in winter, but Limpopo's warmer and wetter winters mean that they stay active for longer.

Why are these weather changes happening? It is because of **climate change**.

Climate Change

Human actions have caused a rapid rise in global temperatures. People have burned **fossil fuels** such as coal, oil, and gas to make energy for over two hundred years. When they burn, these fuels release a gas called **carbon dioxide**. Carbon dioxide traps extra heat on the planet.

Cutting down forests also releases carbon dioxide. Every tree absorbs carbon dioxide

Can you find the climate change effects that are mentioned in this book?

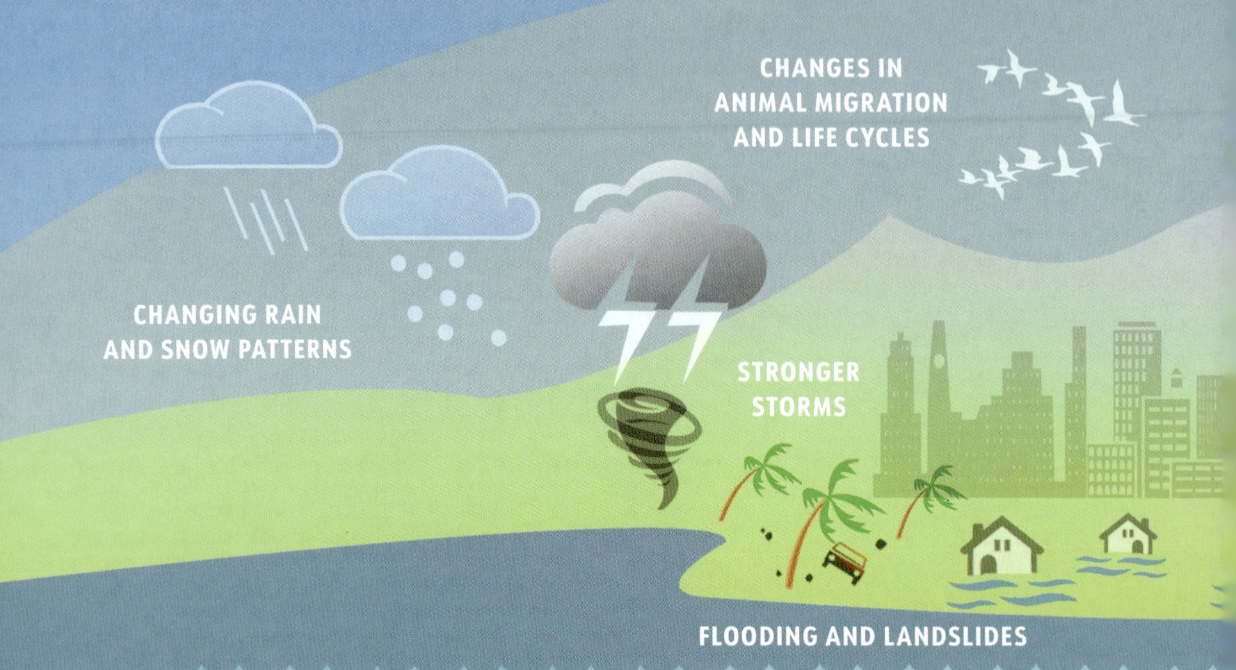

as it grows. Trees help to keep the level of carbon dioxide in proper balance.

This rise in global temperature is changing **weather** patterns all over the world. This is called climate change. Many plants, animals and humans are struggling because of these new weather patterns.

Climate change impacts the well-being of everything on our planet. The effect on Limpopo Province's malaria rates is just one example.

Because of rising temperatures, malaria is spreading to new places across Africa. In areas where malaria already exists, it is challenging

Not just malaria
Climate change is also making Limpopo more at risk of extreme weather events. For example, Cyclone Eloise caused severe flooding in the province in January 2021

Cyclone Eloise, as seen from Space on 22 January, 2021.

efforts to try to control the disease. Africa's goal to eliminate malaria is 2030. Climate change is making this goal even harder to reach.

Our planet is in trouble. We humans need to work together to find ways to slow down the rate of climate change before it is too late.

The flooding caused by Cyclone Eloise created an even wetter environment for mosquitoes to spread.

What Can We Do?

Environmentalists have been warning about the dangers of climate change for a long time. They advise that the best way we can help protect our planet is to "think globally, but act locally".

Water is precious. Climate change is causing droughts in many parts of the world. Use only what you need, then turn off the tap.

Wildfires destroy land and lives, and add to global warming. If you see a wildfire starting, tell an adult immediately.

Every one of us can play our own small part to help our planet. This means that, as a young person, every little step you take adds something good for our shared future.

Avoid buying things you don't need. Instead reuse, look after and repair items instead of throwing them away. This reduces the need to keep burning fossil fuels and polluting the atmosphere.

Find out as much as you can about climate change. Read. Work with your friends. Share ideas for what we can do to make a difference.

Glossary

Anopheles mosquito: The kind of mosquito that infects humans with malaria

Carbon Dioxide: Often referred to by its formula CO_2, this gas traps the heat from the sun in the Earth's atmosphere

Climate Change: The long-term changes in the Earth's weather pattern

Eliminate: To get rid of something

Equator: An imaginary line around the Earth, halfway between the north and south poles

Extend: To make longer, wider or bigger

Fossil fuels: Fuels that come from old life forms that decompose over a long period of time, like coal, petroleum, and natural gas

Hotspot: A place where there is a lot of activity or danger

Malaria: A disease caused by a parasite and passed to humans by Anopheles mosquitoes

Metamorphosis: The process by which most insects, and some other types of animals, such as frogs, change into another stage in their development

Parasite: A plant or animal that lives on or inside another plant or animal and gets its food from it

Repellent: A substance that keeps insects or other animals away

Thrive: To do well, to become strong and healthy

Transmission: The process of passing (transmitting) something on from one person or thing to another

Weather: Natural conditions of a place at one time, including temperature, sunlight, rain, clouds, and snow